YOUR KNOWLEDGE HAS VALUE

Bibliographic information published by the German National Library:

The German National Library lists this publication in the National Bibliography; detailed bibliographic data are available on the Internet at http://dnb.dnb.de .

Imprint:

Copyright © 2016 GRIN Verlag, Open Publishing GmbH
Print and binding: Books on Demand GmbH, Norderstedt Germany
ISBN: 9783668276116

This book at GRIN:

http://www.grin.com/en/e-book/337841/introduction-to-oil-and-gas-environment-the-management-structure-of-bp

Selina Kolls

Introduction to Oil and Gas Environment. The Management Structure of BP

GRIN Publishing

GRIN - Your knowledge has value

Since its foundation in 1998, GRIN has specialized in publishing academic texts by students, college teachers and other academics as e-book and printed book. The website www.grin.com is an ideal platform for presenting term papers, final papers, scientific essays, dissertations and specialist books.

Visit us on the internet:

http://www.grin.com/

http://www.facebook.com/grincom

http://www.twitter.com/grin_com

BP

Management structures

BP (British Petroleum) is one of the largest and most dominant oil and gas companies in the world by market capitalization and revenue. Its business in the energy industry included exploration of fossil fuels, production, and refining, petrochemicals, distribution and marketing. Although its primary dealings are in the fossil fuels sector, the company has recently developed interests in renewable sources of energy such as wind power and biofuels. The corporation, whose headquarters are located in London, is rated among the 'supermajor' in the global energy sector (Hilyard 2012). BP was founded in the early 20th century as the Anglo Persian Oil Company. It underwent several challenges in the first half of the century before it was acquired by the United Kingdom government, which led to a change of name to British Petroleum (BP). Today, it has a presence throughout the world. BP is mainly present in geographical locations where fossil fuels exist as well as the strategic location for refining, transporting and marketing its products (Bamberg 2009).

The most important resource in any organization is the human capital. The success of a business is dependent on the ability of the corporate structures to bring people with shared values together to work towards a common goal. The structure defines how the individuals are built up into a working group. BP is a huge corporation with complex and diverse activities. The management structures are essential in ensuring that the massive workforce, distributed throughout the world, works towards a common objective. In the management structure, people are grouped into layers. Each stratum in the corporate hierarchy has a different level of roles and responsibilities (Albrechtsen & Besnard 2013).

The central decision-making organ in BP operations is the board. According to the company website, the board has the responsibility of proving direction as well as oversight over all activities of the organization on behalf of the shareholders. The board represents the

1

owners of the company and, therefore, is accountable to them. It is composed of thirteen members, which includes the chairman (Carl Henric Svanberg), two executive members, nine non-executive members and one independent director. The executive, through the chief operating officer and the chief financial officer, reports directly to the board. To achieve its mandate, it has established principles and values that guide the organization. The foundation of the principles governing its operations is the importance of "clarity of roles and responsibilities, and the proper utilization of distinct skills and processes". Through a governance report to the shareholders and other stakeholders, the board outlines the performance of the organization as well as the management activities (BP 2016a).

The executive management is composed of the chairman (Carl Henric Svanberg), two executive directors (chief executive officer and chief financial officer) and other executive officers responsible for different divisions and operations in the company. The CEO of BP is Bob Dubley while the chief financial officer is Dr. Brian Gilvary. The executive has the responsibility of managing the day to day operations of the corporation under the supervision of the board. The owners, therefore, confer their authorities to the executive through the board (BP 2016a).

Divisions

The corporate aim of BP is to provide a steady and reliable supply of energy products and services to its clients. To achieve this objective, the company has two major operating divisions, Upstream and Downstream (BP 2016a). The Upstream segment is responsible for finding fossil fuels, extracts through drilling, and transportation. The Downstream unit is in charge of the manufacturing and blending process as well as marketing the BP products. Due to the increased demand for renewable energy, BP has created a third division which deals with investments and development of new sources of energy such as biofuels and wind farms (Ferrier and Fursenko 2016).

Upstream division

The upstream sector forms the foundations of BP operations. It primary responsibilities include exploration of oil and gas in different parts of the world, development of the fields and production. It is also responsible for processes such as transportation, and storage of raw and associated products. According to the BP annual report 2015, the upstream segment of BP operations has five main operational and technical functions. They include exploration function, development reservoirs, as well as global wells, projects, and operational organization. These functions ensure that the resources base of the company is renewed, stewardship in the management of the portfolios, and safety and compliance of operations with internal and external standards (BP 2016b).

The activities of the upstream segment of BP operations are concentrated in 12 regions around the world. They are optimized by a well developed and specialized team of experts in the exploration geology and petrochemical engineering, logistics, procurement and supply chain management as well as financial management. One of the most phenomenon technology used in the exploration of gas and oil is seismic imaging which has played a significant role in supporting the strategic plan of the upstream division (BP 2016b). Despite the immense capital and advanced technology that is involved in this segment of BP operations, maintaining key portfolios in major fossil fuels basins is a daunting task. It has a direct impact on the competitiveness of the group. There are several ways through which the segment competes with other giants in the energy industry. The operations of the upstream division are dependent on its ability to carry out successful exploration and execute projects efficiently, putting into consideration the safety and reliability of the projects. Therefore, it competes with other companies in the sector through operational dependability, safety performance, and access to new oil and gas resources in the traditional and new regions (Ulph 2011).

Downstream division

The downstream segment of BP operations is responsible for manufacturing and blending processes as well as marketing of the company's product. BP deals with three broad categories of product, which include fuels, lubricants, and petrochemicals. The downstream divisions have the responsibility of ensuring that what is extracted from the fields is converted into one of the products marketed by BP. In the fuel subsector, downstream is responsible for refining of the crude product into different forms of fuels, mainly jet fuel, gasoline, diesel and gas, and ensuring that they are available where the customer needs them (Bamberg 2013). Consequently, it is in charge of the management of the supply chain, retail business, and trading activities in the organization. The lubricant subsector involves value addition processes, branding, and marketing of lubricating materials. Over the years, the downstream segment has established partnerships with equipment manufacturers to provide customized and branded lubricants in the market. Petrochemicals are products derived from fossil fuels during the refining process for industrial use. The company has invested in advanced technologies that ensure that they can supply petrochemicals demanded by industrial consumers, mainly paints, plastics and textile industries. This is a critical role because all other aspects of BP business depend on whether it has a market for its products (BP 2016b).

It is essential to note that fuel products in the market are not differentiated. However, the downstream segment in BP has invested in modern advanced technologies which enable the company to remain competitive in the market. The company has branded and customized

4

products, mainly lubricants but can be substituted with products from other firms. However, the principal product, fuel, is not differentiated. The top priorities in the downstream strategy are to adopt technologies that result in advantaged manufacturing, increased efficiency and safe operations. The state of the art refining technologies is the main competitive advantage enjoyed by BP (BP 2016b). Also, the company uses its resources power to finance marketing and sales promotion activities throughout the world. As a direct impact of trademark publicity initiatives, BP is one of the most recognized brands in the global energy market. Additionally, downstream has a well-developed distribution infrastructure and logistic system. They include modern pipeline systems, rail and road transport tankers and sophisticated storage facilities. The division recognizes the significant role of the retail market in BP operations. Consequently, the marketing strategy involves partnering with various retailers to sell its products in the highly competitive market (Herkenhoff 2014).

Contribution to the local community

BP upstream, downstream, and renewable facilities are located in at least seventy countries around the world. Additionally, their products are marketed through different channels in the energy, lubricant and petrochemical market, throughout the world (Bamberg, 2009). Consequently, there are several issues in the company operations that have a direct impact on the local community and environment. While maximizing the return on investment to the shareholders and meeting the demands of its clients, the company has strived to be a responsible and the ideal corporate citizen. Being one of the largest players in the fuel and petrochemical industry, BP has a huge responsibility in ensuring that the global society meets the increased demand for energy. However, there are numerous issues related to the impacts of fossil fuels on the environment that have effects on its operations at all levels (Ulph 2011).

The largest contributions of the BP businesses in the local communities are dependent on the commitments of the organization in promoting ethical human resources management practices. Like other large corporation, BP operations provide employment to the local communities. The locals are employed directly by BP or by its contractors. The company has established a code of conduct that guides the organization on how it relates to its employees. This includes safety in the workplaces, and value for its people. According to the 2015 sustainability report, the performance of the company is dependent on "having a highly skilled, motivated and talented workforce that reflects the diversity of the societies in which we operate" (BP 2016c). The foremost goal of the human resources department at BP is to build capacity in the workforce derived from the host communities. Consequently, the company aims at attracting and retaining the most experienced and talented people. It also provides expanded opportunities for its workforce to build capacity as well as adequately rewarding performance. Other initiatives include promoting health and safety, inclusivity of diverse cultures and engagement of employees. These activities are based on the fact that the company identified its massive human capital as the most important community that contributes to its success (BP 2016c).

In addition to the provision of employment opportunities, there are other social and environmental activities that are undertaken by the company. Some of these activities, although an element of the corporate social responsibility, they are part of marketing strategies. This is because they have direct impacts on the BP brand name. In the recent past, the company has been faced with an uphill task of dealing with the effects of the Deepwater Horizon Oil Spill in the Gulf of Mexico, which has impacted negatively on the reputation and safety of its operations (Partlett and Weaver 2011). Through the operating management system (OMS), BP has developed measures that facilitate the management of social and environmental impacts of all its operations (Gannon 2012). The company has also reiterated its commitment to contributing to the low carbon future. Other environmental initiatives include preparedness for oil spills and related accidents, collaboration in dealing with climate change and greenhouse gasses emissions, water management and unethical practices such as hydraulic fracturing (Ferrier and Fursenko 2016).

The company's operations have both positive and negative impacts on the local communities. They create employment, generate local government revenues and provide opportunities for other business organizations. On the other hand, the activities can result in adverse effects such as destruction of natural environments and cultural heritage (Ulph 2011). Consequently, the company has adopted community engagement strategies through which the local communities can communicate their grievances. While employing the locals, the company invests in local developments programs such as education, disaster relief, donations to charity through the BP Foundation, and working with local indigenous people to promote their culture. These initiatives play a critical role in promoting the BP brand name, and this an important aspect of marketing strategy (BP 2016b).

References

Albrechtsen, E. & Besnard, D. (2013). *Oil and gas, technology and humans: assessing the human factors of technological change.* Burlington, VT: Ashgate.

Bamberg, J. (2009). *The History of the British Petroleum Company.* Cambridge: Cambridge University Press.

Bamberg, J. (2013). *British Petroleum and global oil: 1950-1975: the challenge of nationalism.* Cambridge: Cambridge Univ. Press.

BP (2016a). bp Global. http://www.bp.com/en/global/corporate/about-bp.html

BP (2016b). Strategic Report 2015, http://www.bp.com/content/dam/bp/pdf/investors/bp-strategic-report-2015.pdf

BP (2016c).*Sustainability Report 2015.*

https://www.bp.com/content/dam/bp/pdf/sustainability/group-reports/bp-sustainability-report-2015.pdf

Ferrier, R. W. and Fursenko, A. (2016). *Oil in the World Economy.* Routledge, ISBN 1317234960.

Gannon, M. (2012). *Dissecting a Disaster: The Deepwater Horizon Oil Spill.* Waco: Baylor University.

Herkenhoff, L. (2014). *A profile of the oil and gas industry: resources, market forces, geopolitics, and technology.* New York: Business Expert Press.

Hilyard, J. (2012). *The Oil & Gas Industry: A Nontechnical Guide.* Tulsa, Okla.: Penn Well.

Partlett, D. F., and Weaver, R. L. (2011). B*P Oil Spill: Compensation, Agency Costs, and Restitution.* Washington and Lee Review, 1342-1355.

Ulph, C. (2011). *PR Analysis of British Petroleum.* München GRIN Verlag GmbH.